T0275689

Mobile Data Loss

Mobile Data Loss

Threats and Countermeasures

Michael T. Raggo

AMSTERDAM • BOSTON • HEIDELBERG • LONDON
NEW YORK • OXFORD • PARIS • SAN DIEGO
SAN FRANCISCO • SINGAPORE • SYDNEY • TOKYO

Syngress is an imprint of Elsevier

SYNGRESS.

Syngress is an imprint of Elsevier
225 Wyman Street, Waltham, MA 02451, USA

ISBN: 978-0-12-802864-3

Library of Congress Cataloging-in-Publication Data
A catalog record for this book is available from the Library of Congress

British Library Cataloguing-in-Publication Data
A catalogue record for this book is available from the British Library

For Information on all Syngress publications
visit our website at http://store.elsevier.com/Syngress

DEDICATION

Dedicated in memory of my father Joe, George Morgan (saving each other on 9/11), Michael @hackerjoe Hamelin, Maxx Redwine, Ronnie James Dio, and Gary Eager (you re an inspiration to all of us), all who have dedicated their lives in protecting our country, and all of the loved ones lost on 9/11, my research will forever be dedicated in memory of all of you.

CONTENTS

ACKNOWLEDGMENTS

Mike would like to thank the following people and organizations for their inspiration, mentorship, friendship, faith, motivation, guidance, and support: Coach Konopka (Knops!), Warren Bartley/BSA, Gabe Deale, Randy Paulk, Jim Christy, Domingo Guerra, Roy Wood, Richard Rushing, Black Hat, DEF CON, Wall of Sheep, ISSA

Special thanks to Chet Hosmer, Ojas Rege, John "JKD" Donnelly, Ajay Mishra, and Tom Chang.

And very special thanks to my wife, daughter, and mom for their unwavering support.

CHAPTER 1

Preparing for Generation Mobile

INTRODUCTION

The world is shifting to a mobile-first generation of mobile devices and away from the personal computer. Its impact on our personal lives and our workday is like no other phenomenon of computing history. This phenomenon has brought about a huge shift in how we share and consume information. It has also blended our personal life with our work life. While some companies have attempted to strong arm the wave in opposition, others have embraced mobile thereby allowing these companies to offer solutions that meet the needs and the demand of Generation Mobile.

This tidal wave of mobile devices has been impossible to stop. Whether enterprises embrace them or not, users are accessing your data with mobile devices and sharing it in countless ways, a phenomenon known as Shadow IT. Controlling this data loss may at first appear to be an impossible task. In many cases, users are bringing in their own device and have full control of that device and the data on it. Even those organizations that have issued corporate devices find that in most cases users still find a way to use them for personal use, commonly referred to as COPE (Corporate Owned Personally Enabled).

This has ushered in a new generation of Enterprise IT, one that embraces the value that mobile brings to the company and its business. It's about enablement, agility, disruption, speed, ease-of-use, and abandoning our old approaches to IT. But this doesn't mean that security is an afterthought. We must learn from the security flaws of the PC era, embrace the enhanced security that mobile provides, and take our enterprises securely into the next generation of computing—mobile computing.

THE PROBLEM

More than 1 Billion data records were exposed in 2014 alone[1] with legacy PCs and Servers as the main target.[2] These records include social security numbers, home addresses, credit cards, health information, fingerprints, and much more. The exposure has been global, encompassing virtually every industry and governments across the globe. It's quite a staggering statistic considering that we've had roughly 30 years to perfect and fortify these PC-era technologies, and it appears that we're worse off than we've ever been.

These legacy PC and Server-based operating systems rely on security products such as anti-virus to provide an overlay agent that encompasses a firewall, anti-malware, anti-virus, and intrusion prevention. Add in additional network layers that include a perimeter Firewall, Network Intrusion Prevention, Malware Protection Systems, Proxies, and more. Yet as evidenced by the mass breaches of 2014 and 2015, we're losing the battle. Malware continues to infest these systems, and attackers are winning the battle.

A few lessons can be learned from these breaches. The first is that the length of time to discover a breach ranges from weeks to months, or even longer. Roughly 50% of breaches take months to detect.[3] This would imply that there is a massive time window of compromise stemming from a huge lack of visibility and automated timely countermeasures to these attacks to mitigate a breach.

Secondly, less than 4% of the breaches from 2014 "involved data that was encrypted in part or in full."[4] With the widespread availability of encryption solutions, it's clear that very few organizations have embraced encryption or deployed it to protect their most important data.

---

[1]2014 Year of the Mega Breaches & Identity Theft, http://breachlevelindex.com/pdf/Breach-Level-Index-Annual-Report-2014.pdf
[2]2014 Data Breach Investigations Report – Verizon, http://www.verizonenterprise.com/DBIR/2014/reports/rp_dbir-2014-executive-summary_en_xg.pdf
[3]2014 Data Breach Investigations Report – Verizon, http://www.verizonenterprise.com/DBIR/2014/reports/rp_dbir-2014-executive-summary_en_xg.pdf
[4]Breach Level Index Annual Report 2014, http://breachlevelindex.com/pdf/Breach-Level-Index-Annual-Report-2014.pdf

Lastly, many enterprises rely on perimeter security technologies as the foundation of their security strategy.[5] As mobile becomes the primary computing method, these enterprises are ill-equipped to protect their data from ubiquitous mobile devices, cloud services, social media, and the new generation of mobile computing. Mobile doesn't respect traditional network boundaries. And many of these organizations see mobile security solutions as a bolt-on product to their perimeter security strategy, rather than a fundamental shift in how enterprise data is shared and protected.

WHAT'S DIFFERENT ABOUT MOBILE?

Managing and controlling data-at-rest on a legacy PC is difficult. The operating system provides very little in terms of isolating corporate data from personal. And in most cases all applications have access to all data on the PC. If you have access to the PC, you're considered a trusted user. This provides a huge threat surface leading to data loss, malware attacks, and breaches (Figure 1.1).

Mobile operating systems are different from their PC counterparts in that they employ operating system sandboxing. This sandbox approach separates each app and its data from other apps and their data. This also

Open file architecture – All apps
have access to all data

Sandboxing – Each app
and data isolated from
one another (containers)

Figure 1.1 Operating systems – PC vs. Mobile.

[5]Breach Level Index Annual Report 2014, http://breachlevelindex.com/pdf/Breach-Level-Index-Annual-Report-2014.pdf

includes isolation from the operating system as well. But there are features in the mobile operating systems that provide ways in which data can be shared and are typically user-driven. A user can receive an email with an attachment in the email app, open that attachment in a secondary app that allows for it to be edited, and then open the document in a third app to print it over-the-air to a printer, and furthermore upload it to a cloud service. Additionally, features like copy/paste, screenshot, email forwarding, and more exist as well. But what's important is that much of this is user-driven or user-defined rather than allowing an app to natively perform these functions.

Another important aspect of the mobile era is that the traditional network edge has now become blurred. Mobile devices are very ubiquitous and access enterprise data over the network in a variety of ways. Whether its cloud services, web 2.0, data backup services, multiple network services (cellular, Wi-Fi, NFC, etc.); all make management of this data far more challenging. No longer can we look at the network as a single entry point, the network edge has disappeared, now data lives everywhere.

Last, but certainly not least, is the emergence of BYOD (Bring your own device). In the PC world, IT provided the computer preconfigured with security controls. But in the mobile world, people show up with their personal devices looking to connect them to their enterprise network or cloud. And even those organizations with Corporate issued devices, inevitably find that the user will use it for personal use. In either circumstance, the user has a plethora of features by which they can share, forward, or upload data to and from the network. This has also made the end-user the low hanging fruit for attack. Since these mobile devices are always connected, this provides a much larger window of compromise for attack and exfiltration of data.

SECRETS TO MOBILE DATA LOSS PREVENTION SUCCESS

Mobile data loss requires a mindset adjustment to enterprise security that adapts to the mobile operating systems and explosion in data dissemination. Enterprise Mobility Management (EMM) is the centerpiece for accomplishing this. Apps, content, network access, and email deployed to the mobile device can be revoked through a selective wipe when a device is lost or stolen, or if the device is identified as compromised. Furthermore, the EMM can work in unison with a secure mobile gateway to block access for devices that are unknown or identified as out-of-compliance. Without EMM, your organization will be crippled by lack of visibility, control, and automation to respond to risks and threats.

EMM also compliments the aforementioned detect-and-respond techniques with greatly enhanced *defensive* techniques to *proactively* prevent data loss, breaches, cyber espionage, malware infestations, and network threats. This is accomplished through the use of data loss prevention (DLP) controls, encryption, containerization, and much more. It's safe to say that most organizations would rather avoid a data loss breach than have to respond to one. EMM allows the organization to secure devices, apps, content, network access, and cloud. But success of the program is not all about lockdowns, it also involves providing ease-of-access to the users, and protecting their privacy.

In mobile, users are accustomed to ease-of-access to content as a result of an apps-based model. One-click access allows access to their email, calendar, social media, cloud repository, and much more. Balancing this with security takes a careful approach that makes the security more invisible to the user and gives the look and feel of the native experience they are accustomed to. This can be accomplished by using mobile-specific approaches such as app-based security, certificates, per-app VPN, and other technologies commonly used in mobile.

Balancing mobile security with privacy is the sword by which your security strategy will survive or fail. At a time when people share everything about their lives in social media, they are paranoid about what a company can see or not see on their mobile devices. Paranoia about this has increased immensely from the PC era to the Mobile era. Surprisingly, there's very little that a company can see on a mobile device that they're managing. In 2015, a "trust gap" survey was performed to understand the variance in what employees thought an employer could see on their mobile device versus the very limited personal information that an employer can actually see.[6] Contrary to popular belief, employees cannot see personal emails, attachments, text messages, photos, videos, voicemails, or personal web browsing activity. Remaining transparent to your employees about this is crucial to your security strategy's success. It can be outlined in policies displayed when they register their device, as well as in a management app with visual privacy through which they can view this anytime. Taken from a company with a hugely successful mobile security program "We do not want to see what you do, but make what you do safer."

SUMMARY

In summary, an enterprise mobile security strategy should encompass the User, Apps, Device, and Network. This goes beyond the device and beyond EMM, and encompasses a holistic mobile security design. In the next chapter we'll dive deeper into the mobile threat landscape and the impact of these threats on enterprise mobile data loss. This will provide the basis for outlining countermeasures for existing threats as well as future ones, and provide a baseline for protecting enterprise mobile data.

It's important to note that your enterprise mobile security strategy will not become a dusty security policy meant to cover mobile security for the next 10 years. Instead, your mobile enterprise security strategy will develop into an iterative security framework that requires analysis and refinement over time. The goal is a more *defensive* approach to protect your enterprise data with the enhanced security and features provided by mobile, and incorporate lessons learned from the plethora of PC-related breaches of the last few years. Mobile has introduced more secure options to protect your enterprise data; features that adapt to the new world of data sharing. The risks and threats may be evolving, but so are the countermeasures. Welcome to the mobile era of security.

CHAPTER 2

Understanding Mobile Data Loss Threats

MOBILE THREAT VECTORS

In the last chapter I outlined three main differences in mobile versus the PC era:

- Mobile operating systems leverage sandboxing techniques to isolate apps and their data from one another.
- The network perimeter has become blurred; data now lives everywhere, on the device, on the network, in the cloud, and within apps.
- BYOD has ushered in an opposite approach for IT; users own the device, but IT wants to ensure enterprise data is secure on the device while maintaining the user's privacy.

This changes the threat landscape and creates a new attack surface for attackers. This impacts how we protect against malware, data risks, network attacks, and compromises. Let's dig deeper into the mobile threat vectors.

MOBILE OS COMPROMISE

Users will commonly jailbreak (iOS) or root (Android) a device to customize their device, and unlock additional features and functionality. This typically involves connecting their device to their PC or Mac and using freely available software designed to perform the jailbreak or rooting activity. This jailbreak or root activity may also stem from malware that exploits vulnerability in the mobile operating system. Most are not aware of the security implications of a jailbreak or root.

A jailbreak or root on the device will unlock additionally functionality, services, and the ability for the user to download apps outside of the curated app stores. As a result, the security becomes greatly diminished, making the device much more vulnerable to malware, privilege escalation, network attacks, and ultimately data loss.

An iOS jailbreak typically requires that the attacker enter their PIN and then pair the device to their PC or Mac with iTunes by confirming the "trust" notification. Without the PIN (assuming it has one), you cannot pair the device to a PC or Mac, diminishing the ability to jailbreak a device. This is one of the many reasons why a PIN or Passcode is fundamentally important to protect against someone other than the user from accessing data on it by jailbreaking the device. This presents a challenge to the attacker as most MDM/EMM (Mobile Device Management/Enterprise Mobility Management) products enforce a PIN or Passcode with an automated policy to wipe the device after 10 failed login attempts.

On Android, there are many variants of rooting. In fact, some devices come from the factory already rooted! These devices may come with a custom ROM or backup software installed that requires the device to be rooted in order for backup software to function. Other threats can stem from a user who enables the ADB (Android Debug Bridge) or USB controls to tether it to a PC or Mac and side-load an app outside of the Google Play. All of these can lead to a compromise of the Android device leading to escalation of privileges, weakening of root permissions, and other indications of a compromise.

Device manufacturers, carriers, and others modify the Android OS leading to a plethora of Android variants, and unknowingly can create vulnerability in their build. There have been many Android operating system compromises through malicious apps that expose a device vulnerability to allow privilege escalation.

Here's a short list of various mobile operating system compromises and risks:

- Jailbreak (variants including Pangu and Evasion)
- XCon (Jailbreak anti-detection)
- Rooting (variants)
- Android ADB/USB Controls
- Android Custom ROMs
- Android Modified file permissions

From a user or administrator standpoint, PIN and encryption is the first line of defense that can be used to mitigate the threat of attack. In a lost or stolen scenario, the attacker must typically first bypass the

PIN protection in order to pair the device to a PC or Mac before the jailbreak or rooting can be performed. Considering the broad range of mobile operating system compromises, detection capabilities are also critically important to the management ability to quarantine devices.

MALWARE AND RISKY APPS

The PC world is inflicted with malware from a myriad of different attack vectors ranging from the operating system vulnerabilities, network services, applications, middleware, browsers, and more. The lack of application sandboxing makes the average PC vulnerable to file infections that impact the operating system, applications, and data. The operating is vulnerable to file infections from malicious websites that prompt a file download to the PC, email attachments with infected file attachments, viruses passed on by a shared USB drive, and a plethora of other threat scenarios. Furthermore, these legacy operating systems have little-to-no kernel protection. Lastly, users are not provided any form of a curated app store. Aside from business apps, other applications are downloaded from millions of websites with no known integrity or trust. The result is that a file can impact all apps, all data, and typically the operating system.

This is somewhat different from mobile malware, which is typically deployed through apps rather than files. This can stem from a phishing attempt through a malicious email or SMS text message that prompts the user to download an app or an update to an app. In addition, malicious apps sometimes make their way into one of the curated app stores, such as with the XcodeGhost malware. More on this is discussed in the next chapter.

As demonstrated by the Masque Attack[1] on iOS devices, threats such as these don't require a compromise of the mobile operating system, such as a Jailbreak for iOS or Rooting of an Android device. Instead the user is coerced through the phishing attack to download an update to an existing app. Since this app is signed using the enterprise signing certificate, the updated app is considered trusted and is therefore updated and now infected with malware. This malware can steal

---

[1]https://www.fireeye.com/blog/threat-research/2014/11/masque-attack-all-your-ios-apps-belong-to-us.html

Anatomy of the Masque Attack (iOS)

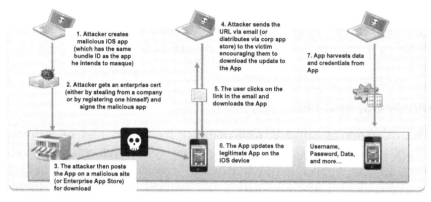

1. Attacker creates malicious iOS app (which has the same bundle ID as the app he intends to masque)

2. Attacker gets an enterprise cert (either by stealing from a company or by registering one himself) and signs the malicious app

3. The attacker then posts the App on a malicious site (or Enterprise App Store) for download

4. Attacker sends the URL via email (or distributes via corp app store) to the victim encouraging them to download the update to the App

5. The user clicks on the link in the email and downloads the App

6. The App updates the legitimate App on the iOS device

7. App harvests data and credentials from App

Username, Password, Data, and more...

Figure 2.1 Masque Attack.

data from the app, as well the app credentials. Figure 2.1 outlines the attack.

Stagefright[2] was a vulnerability found in the media library on Android that impacted approximately 99% of all Android devices. An attacker can send a malicious multimedia message via MMS. When a vulnerable Android device receives message, it automatically downloads (default setting) and infects the device through the multimedia preview function. This can allow an attacker to steal data, hijack the microphone, use the camera, and essentially behave like spyware on the infected device. The fragmentation with Android presents a challenge when attempting to patch Android devices. Unlike Apple's iOS where all patches come from Apple, Android relies on carriers to provide the patches to their respective Android devices. Many times patches are delayed for months, and in other cases never provided.

Risky apps are another concern that can present a risk to enterprise data. Apps that collect location information, harvest contacts, collect device hardware information, and more may not directly present a malware threat, but do present a privacy risk to user and enterprise

[2]https://blog.zimperium.com/stagefright-vulnerability-details-stagefright-detector-tool-released/
Stagefright: Vulnerability Details

Figure 2.2 Wireshark packet capture of an App communicating clear-text with no encryption.

data. And many of these fall into the realm of Personally Identified Information (PII) as outlined by NIST 800-53[3] publication.

Developers may not always be focused on security and therefore employ poor coding practices such as not encrypting data-at-rest or data-in-motion. This may not only expose the data, but also credentials such as the user's username and password, web tokens, keys, and other sensitive information.

Free apps commonly are designed to monetize the app through marketing, adware, or redirect to an alternative search engine. Here's an example of an app that claimed to be using encryption for the data-in-motion, but is sending the password clear-text as demonstrated by the Wireshark packet capture (Figure 2.2):

Also, app developers commonly use SDKs and libraries from untrusted Internet sites and use them in their apps. This can cause them to unknowingly embed malware, adware, and other risky components into their apps. A good example of this was XcodeGhost. Xcode is the SDK provided by Apple to developers to be used for creating apps for iOS. But many developers download Xcode from websites other than Apple. Some of these sites were found to have malware infested versions of Xcode, known as XcodeGhost. Developers unknowingly created apps, or updates to apps, using these infected versions of Xcode. As a result, it was found that XcodeGhost infected apps were uploaded to the Apple App Store and downloaded by innocent users who subsequently infected their devices.

In summary, attackers have shifted their focus to apps thereby making this the most common mobile threat vectors. Malicious apps and risky legitimate apps must always be considered when mapping out the mobile threat landscape and risks to enterprise data.

---

[3]http://nvlpubs.nist.gov/nistpubs/SpecialPublications/NIST.SP.800-53r4.pdf Security and Privacy Controls for Federal Information Systems and Organizations

USER DATA LOSS

Users are empowered with a plethora of ways on a mobile device in which data can be shared across apps, to cloud repositories, device backup syncing with cloud services, and more. Some data loss is accidental, while others are intentional. The 2015 Verizon Data Breach Investigations Report concluded that 20.6% of incidents were caused by insider misuse.[4] And an additional 15.3% stemmed from physical theft or loss.

Many inherent features are available in iOS and Android to allow data sharing. In iOS, features such as AirDrop and URL Schemes allow sharing of data.[5] In Android these are referred to as "intents.[6]" These features, while respecting the application data sandboxing, do allow some user-driven behaviors. For example, an user can receive a corporate email with an attachment in their email app, open the attachment in a document editing app, and then upload it to a personal cloud file and sync sharing service through a third app. Additionally, the mobile operating systems allow features like copy/paste, open-in, upload, screenshot, and more. The real challenge to the mobile administrator is that these behaviors change every time a new version of the operating system is released, and releases are frequent!

Many of these user-driven data loss threats are accidental. Our mobile devices are designed to automatically sync with backup cloud services. Users are frequently unaware that they may be backing up sensitive corporate data to their personal cloud services.

User data loss can also occur from the myriad of network services running on a device. These can include cellular, Wi-Fi, Bluetooth, NFC, IRDA, and more.

Cyber espionage is also growing at an alarming rate. The implications in mobile are huge. In fact, the Verizon Data Breach Investigations Report 2015 concluded that cyber espionage compromised 18% of "confirmed data breaches."[7] Users can share data in a variety of ways, as mentioned earlier. But IT management of mobile devices through an Enterprise Mobility Management solution may

---

[4]http://www.verizonenterprise.com/DBIR/2015/ Data Breach Investigations Report by Verizon
[5]https://developer.apple.com/library/ios/documentation/iPhone/Conceptual/
iPhoneOSProgrammingGuide/Inter-AppCommunication/Inter-AppCommunication.html
[6]https://developer.android.com/guide/components/intents-filters.html
[7]http://www.verizonenterprise.com/DBIR/2015/ Data Breach Investigations Report by Verizon

allow a developer, business unit, and other non-IT folks to distribute or update malicious apps to users. The implications are that if one of these individuals has malicious intent, it would be quite easy for him or her to distribute a malicious app that contains spyware that could allow access to SMS messages, email, or even the microphone or video camera to spy on an individual. While many would lump this into malware or malicious apps, they would only be partly correct. The intent and method of infestation are arguably different. In the next chapter, we'll outline an approach to tackle this issue.

Another issue is your productive users. Many users look for ease-of-use when sharing data with other employees, customers, and business partners. If IT takes a stance restricting everything, users will inevitably find a way around it, a phenomenon known as Shadow IT. This can be one of the biggest data threats. Users want to use the cloud to share content; it makes it easier than what IT has provided. But when this occurs, IT looses at visibility and control of the data, and indirectly represents a data breach. Fortunately, enterprise file and sync share (EFSS) solutions exist today to allow secure file and content sync and sharing that allows them to securely store data in the cloud and continue to use their personal cloud shares. More of this will be covered in the next chapter. For now, this is another threat vector to enterprise data to include in the overall list of threats.

MOBILE NETWORK THREATS

We live in a mobile world, and as a result our users are mobile. Users will commonly connect to any free Wi-Fi they can find while they're traveling or away from the office. This includes coffee shops, hotels, airports, and other public places. Attackers know this and as a result locations with Open Wi-Fi are hunting grounds for attackers. Attacks can range from simple interception of unencrypted Wi-Fi traffic to more sophisticated Man-in-the-Middle (MitM) attacks targeted at encrypted traffic.

To perform an interception attack, the attacker can simply associate to the same wireless access point as the target user. Now that network access has been established the attacker can simply start his wireless sniffer (commonly wireshark) and begin capturing traffic looking for unencrypted transmissions such as HTTP, telnet, ftp, and more. This exposes more than account logins and passwords. It also exposes

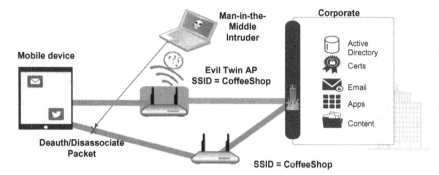

Figure 2.3 Man-in-the-Middle Attack.

URLs with cookies, web tokens, and a plethora of other sensitive information. Many of these URLs can be "replayed" or copied and pasted into their own browser to get authenticated access to user data. This is low hanging fruit for an attacker.

More sophisticated attacks on an Open Wi-Fi network may target encrypted transmissions back to the corporate network. When users want to remotely connect to corporate resources, they frequently use a VPN. Attack tools such as Backtrack, Kali, or a Wi-Fi Pineapple[8] can be used to perform a Man-in-the-Middle attack. An attacker can set up a fake virtual or physical access point with the same SSID name as the legitimate coffee shop access point. Unknowing users may connect to this seemingly innocent access point and proceed to remotely connect to their corporate network over the VPN. The Fake Access Point (AP) can be used to terminate the encryption tunnel, and create a secondary VPN tunnel to the corporate network. By terminating the encryption tunnel at the Fake AP allows the attacker to decrypt the traffic and read it clear-text. In fact, this could also expose the corporate network and allow the attacker to perform further attacks on the corporate environment (Figure 2.3).

Note that these same attacks could be performed on an enterprise guest Wi-Fi network as well! A corporate guest Wi-Fi network is designed for contractors and guests, but is also commonly used by

---

[8]http://hakshop.myshopify.com/products/wifi-pineapple

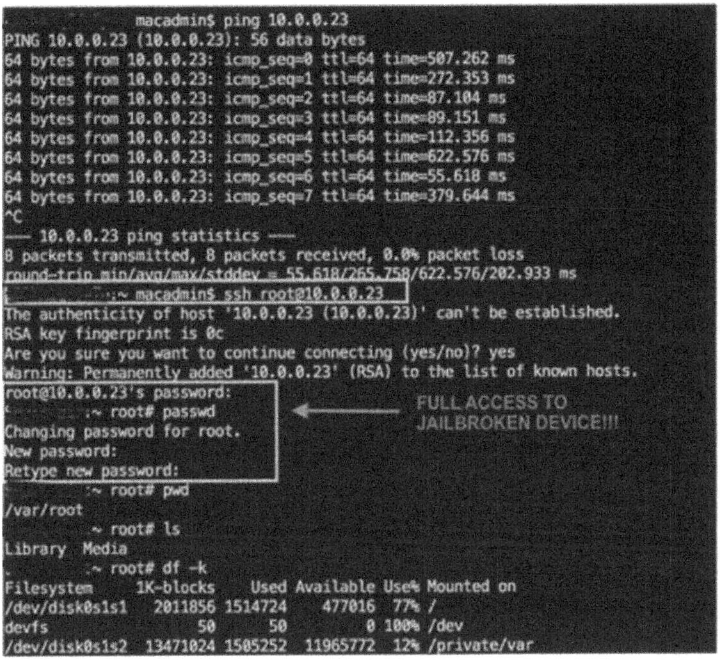

```
                 macadmin$ ping 10.0.0.23
PING 10.0.0.23 (10.0.0.23): 56 data bytes
64 bytes from 10.0.0.23: icmp_seq=0 ttl=64 time=507.262 ms
64 bytes from 10.0.0.23: icmp_seq=1 ttl=64 time=272.353 ms
64 bytes from 10.0.0.23: icmp_seq=2 ttl=64 time=87.104 ms
64 bytes from 10.0.0.23: icmp_seq=3 ttl=64 time=89.151 ms
64 bytes from 10.0.0.23: icmp_seq=4 ttl=64 time=112.356 ms
64 bytes from 10.0.0.23: icmp_seq=5 ttl=64 time=622.576 ms
64 bytes from 10.0.0.23: icmp_seq=6 ttl=64 time=55.618 ms
64 bytes from 10.0.0.23: icmp_seq=7 ttl=64 time=379.644 ms
^C
— 10.0.0.23 ping statistics —
8 packets transmitted, 8 packets received, 0.0% packet loss
round-trip min/avg/max/stddev = 55.618/265.758/622.576/202.933 ms
            :~ macadmin$ ssh root@10.0.0.23
The authenticity of host '10.0.0.23 (10.0.0.23)' can't be established.
RSA key fingerprint is 0c
Are you sure you want to continue connecting (yes/no)? yes
Warning: Permanently added '10.0.0.23' (RSA) to the list of known hosts.
root@10.0.0.23's password:
        :~ root# passwd              <---  FULL ACCESS TO
Changing password for root.               JAILBROKEN DEVICE!!!
New password:
Retype new password:
        :~ root# pwd
/var/root
        ~ root# ls
Library  Media
        :~ root# df -k
Filesystem     1K-blocks     Used Available Use% Mounted on
/dev/disk0s1s1 2011856  1514724   477016  77% /
devfs              50       50        0 100% /dev
/dev/disk0s1s2 13471024 1505252 11965772  12% /private/var
```

Figure 2.4 Access to Jailbroken iOS Device over Encrypted Corporate Wi-Fi Network.

employees to access websites that may be blocked by the internal corporate network. Therefore these same risks should be considered at your facilities as well.

Users of secure Wi-Fi networks at work can also be targeted. Even when AES-256 bit encryption, WPA/WPA2, and more are used, Wi-Fi users can normally communicate with one another over the secure and encrypted Wi-Fi. Unless PSPF (Public Secure Packet Forwarding[9]) or Client Isolation is used on your "secure" Wi-Fi network, a malicious insider can identify a jailbroken iOS device and attempt to log onto it. Most iOS devices ship from the factory with a default username of "root" with a password of "alpine." Unless the user changed that password when they jail broke their device (and rarely do they change the default password), *anyone* on that same network can log into their jailbroken device and gain *full access!* (Figure 2.4).

[9]http://www.cisco.com/web/techdoc/wireless/access_points/online_help/eag/123-02.JA/1400BR/h_ap_network-if_802-11_c.html

SUMMARY

What we've provided is a categorization of mobile threat vectors to provide a baseline for outlining countermeasures. These include:

- Mobile operating system compromises
- Malicious and Risky Apps
- User-driven data loss
- Mobile network threats

Threats will continue to emerge over time. For example, smartwatches, wearables, and the Internet of Things (IoT) are all devices that can pair with mobile devices. Later in the book I'll outline this emerging threat and countermeasures for protecting enterprise data. What's important is to stay on top of those threats by subscribing to threat news feeds, embracing your security vendor notifications, and doing your own research. It's all part of the security lifecycle. In the next chapter, I'll begin to outline countermeasures to these threats and outline a strategy to tackle them from proactive, reactive, and live scenarios.

CHAPTER 3

Mobile Security Countermeasures

So far I've outlined many of the mobile device threats that could lead to data loss. Fundamentally, when considering data loss one must encompass data-at-rest and data-in-motion to ensure confidentiality and integrity of the data. But a mobile device is more sophisticated than that. This involves protecting data on the device, data in the app, and data over the network (Figure 3.1).

Fortunately, mobile devices and complimentary products leverage new features in the mobile operating systems not previously found in traditional PCs. Let's continue by detailing these newer features and outline countermeasures to many of these aforementioned threats.

MOBILE OS COMPROMISE

In the previous chapter I outlined a myriad of ways in which a mobile device can become compromised. There are multiple approaches for detecting and mitigating this threat. First, the EMM client should provide ways to identify an OS compromise locally on the device, and then report that back to the console. In response, the administrator should have a policy to quarantine devices when a compromise is detected. This automation should allow the console to send down a Selective or Full Wipe of the device. A selective wipe would remove the enterprise data only, while leaving the personal data alone. A full wipe of course wipes the entire device back to factory defaults, and is typically only suited for corporate-owned devices. Selective wipes can be accomplished in a few ways. One way is to remove the previously deployed configuration profiles such as email, Wi-Fi, VPN, etc. Additionally, managed apps and/or their data can also be removed (note that this capability varies across the different mobile operating systems). When using a container, the selective wipe would purge the container itself.

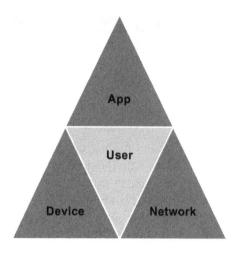

Figure 3.1 Mobile Data Loss Protection Triad.

Also, when a compromised device is detected, other lockdowns can occur. For example, the mobile device can also be automatically blocked from remote access to the network by a secure mobile gateway, until the device is brought back into compliance. The same can be done for the local network. A similar approach can be employed with NAC (Network Access Control), where the NAC solution checks in with the MDM/EMM when a device connects to the network to determine its security posture and if it's a registered device. If out of compliance, the NAC can block access similar to a secure mobile gateway. In terms of cloud services, EMM integration with Azure Active Directory can block rogue and out-of-compliance devices from accessing Office 365.

It's important to note that there's an issue not addressed by the aforementioned countermeasures that is lost or stolen devices. Assuming the lost or stolen device remains on the network, the EMM can still receive threat notifications from the EMM client and issue a quarantine to protect corporate data with a selective wipe. But if the device is a Wi-Fi-only device and it's no longer on the Wi-Fi, how does the EMM still quarantine the device? If it's off the network, the EMM loses visibility into the device.

More recently some EMM products have added *offline* policies that can reside on the device, specifically when using a container solution for your enterprise data. The local EMM client can still look for the same types of OS Compromise threats, but now when a threat is detected it doesn't need to "phone-home" to the EMM management console to receive a quarantine command. Instead a local policy selectively wipes the container. This is particularly helpful in organizations that have many Wi-Fi-only mobile devices. In fact, the PCI Council added this to its Mobile Point-of-Sale (POS) "Mobile Payment Acceptance Security Guidelines v1.1, July, 2014.[1]"

Most recently in Windows 10, the operating system now performs a device health check to validate the integrity of the device during the bootup process. This can then be reported to the MDM or EMM and used to block access to corporate resources.

Summary of Mobile OS Compromise Countermeasures:

- PIN or Password enforcement
- Encryption
- Containerization of enterprise data
- OS Compromise detections (Jailbreak and Root detections) and Quarantine
  - Online selective wipe
  - Offline selective wipe
- Out-of-compliance device triggers the network gateway to block access

MALWARE AND RISKY APPS

Based on the plethora of threats I outlined in chapter "Understanding Mobile Data Loss Threats," it's important to detail an approach to deterring malware and risky app behaviors. Since we know that iOS is no longer immune to malware threats, a comprehensive mobile security strategy should address these threats across all of your mobile devices.

Anti-virus alone has taken a backseat to more comprehensive mobile malware security products. The reason for this is that on a mobile device anti-virus is just another app, and therefore the sandboxing limits its ability to remove a malicious app, limiting it to alert the user and rely on them to remove it. This is very different from the PC world where we've always relied on anti-virus to both identify the threat *and remove it*.

Due to this shortcoming of anti-virus alone, a new group of products has emerged referred to as App Reputation and Mobile Threat Prevention. This is a broad exploding category of products

[1] https://www.pcisecuritystandards.org/security_standards/documents.php

designed for mobile threats. The key difference here is that they all integrate with the EMM to leverage the EMM's ability to respond to an identified threat with a quarantine.

App Reputation commonly uses the EMM app inventory of the mobile devices under management and correlates it against their database of known malicious and risky apps. It will then report on malicious or risky behaviors for each app, either in its own console or also in the EMM console to give the administrator a single monitoring dashboard. The App Reputation may then feed into an EMM App blacklist to spawn a quarantine. It may also tie into APIs to allow profiles to be removed from the device and selectively wipe corporate data.

Mobile Threat Prevention is also a broad category of products that rely largely on an anti-virus-like app on the device that may include some intrusion detection features, malicious app behaviors, and more. These products can also integrate with an EMM to kick off a quarantine when a threat is identified on a mobile device. Furthermore, some of the features between App Reputation vendors and Mobile Threat Prevention vendors have also begun to overlap. Some App Reputation vendors have added an app to analyze local behaviors on the device, thus providing a more defense-in-depth approach.

These products are changing quickly with more features always being added. App Reputation and Mobile Threat Prevention solutions are very important to an overall Mobile Security Strategy as concerns about malware continue to increase.

ACCESS CONTROL AND CONDITIONAL ACCESS

Ensuring the network is secure for remote access is key in a mobile world. Traditionally in the PC world this has been delivered through a remote access VPN. Mobile requires a more mobile aware secure gateway. This gateway can control access to resources such as ActiveSync or Lotus Notes email. In addition, it can control access to content, internal web services, and application servers. Access control is performed by authenticating the user and the device.

When a device is under MDM or EMM management, the management system can collect hardware and software information about the device. This is key to eliminating impersonation and cloned devices, and

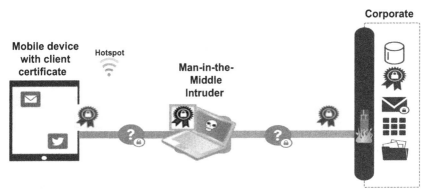

Figure 3.2 Thwarting a Man-in-the-Middle Attack.

used for authenticating the device. In addition, the security posture can be analyzed to identify when a device is outside of corporate compliance policies, as defined in the security policy. By combining this with user authentication, the device authentication provides yet another factor of authentication when a device remotely connects to the network and is far superior to traditional gateways.

Most of the mobile operating systems have native support for certificates, making it quite easy for certificates to be deployed with an EMM profile automatically for authentication, unlike their PC counterparts, which normally required cumbersome manual techniques for deploying certificates to users PCs and laptops. Therefore, when a profile is deployed to a device for services such as email, SharePoint, and intranet web access, a certificate can be generated and deployed to the device automatically. This also eliminates hassles such as required password changes every 90 days. It also allows an organization to meet security or compliance requirements requiring strong factor or two-factor authentication. When combined with a secure mobile gateway, it also provides *proactive* protections against MitM attacks by offering both mutual authentication, and certificate pinning on the secure mobile gateway (Figure 3.2).

Steps to thwarting a MitM attack:

1. Attacker presents fake server-side certificate (impersonating the network back at corporate)
2. Certificate pinning prompts the fake certificate to be compared to what has previously been sent to the device and quickly identifies that they don't match

3. Client certificate mutual authentication handshake fails
4. No per-App VPN tunnel is set up
5. No data communicated
6. Data breach is prevented

A secure mobile gateway also can support mobile-specific encrypted protocols, such as per-App VPN over SSL/TLS. This was released in iOS 7, and gained mass support across public apps in iOS 8 and iOS 9. Supporting a VPN at the app-level allows the administrator to further refine what apps can access the corporate network. In contrast, a VPN typically allows all apps to access the network, including malicious apps. A per-App VPN provides additional layers or security as well as better efficiencies and ease-of-access for the user.

LOCKDOWNS AND RESTRICTIONS

Lockdown and restriction APIs have been available from device manufacturers for some time, and allow EMM solutions to leverage these APIs to disable features. These include unwanted network services (Bluetooth, IRDA, NFC, etc.), device level features (camera, screenshot, etc.), and a plethora of other lockdowns. These vary across the different mobile operating systems.

Furthermore, many EMM solutions allow these to be applied to manage mobile devices in different ways. For example, for a mobile POS, unwanted services such as Bluetooth or NFC can be disabled to avoid targeted attacks. But disabling these on BYOD devices may not be desirable since users commonly use these services for Bluetooth headsets, NFC-based retail purchases, and more. It's important to ensure when implementing these controls to evaluate each of the use-cases and perhaps different lockdown and restriction policies for each scenario.

LIVE MONITORING, AUDIT LOGS, EVENTS, AND REPORTING

EMM solutions provide inherent live monitoring of mobile devices. This can be mobile device monitoring, device security posture monitoring, network access monitoring, and more. Additionally, EMM can integrate with SIEM, Big Data Analytic products, App Reputation, Mobile Threat Prevention, Network Access Control, and proxy solutions. All of these provide the ability for logging, alerting, correlation, and reporting.

The administrator can force a device check-in to check the security posture or location of the device. Per-device logs can be stored in the EMM to allow deep analysis by the administrator. While this may be helpful for troubleshooting, it can also be helpful for security analysis. Furthermore, an EMM can provide information about when a device is connected to a network, and to what resources.

INCIDENT RESPONSE AND FORENSICS

In the event of a breach or incident, investigators are quick to perform an acquisition on a mobile device. But on-device mobile forensics is becoming increasingly difficult. Many of the mobile device forensic acquisition tools have historically required vulnerabilities, hacks, or even formal jailbreak or root to bypass protections to gain access to the data. As previously outlined, many of these techniques may perform a wipe or selective wipe, even if the device is off the network (or in a Faraday bag).

A blind spot for many investigators is that an EMM may hold some significant evidence. It also doesn't require breaking into the mobile device. The following list outlines just a sampling of the EMM data available to the investigator:

- Remote unlock of the mobile device
- Device hardware and software information
- An inventory of Apps on the device
- Last known location of the device or a bread crumb trail of where the device has been
- When the mobile device connected to the corporate email
- When the mobile device connected to corporate app
- What malicious apps are installed on the device that may have led to a breach
- When a malicious app was installed on a device
- If the device is compromised
- When the device was compromised
- Audit logs of files uploaded to personal cloud services

As you can see, an EMM solution provides a wealth of information to the investigator to answers questions such as when, where, what, why, and how. While although on-device forensic acquisition is valuable, EMM may provide answers more quickly and easily. This is especially important in the event of a breach and time-to-resolution; and is incredibly helpful in a liturgical and nonliturgical forensic investigations. Mapping out these EMM data points is key to updating your incident response lifecycle and response procedures.

MOBILE DEVICE UPDATES AND PATCHING

In sharp contrast to the PC world, users are in control of when mobile device updates and patches are applied. This is very much the case with iOS and Android, and is becoming more of the case with Windows such as in Windows Phone 8/8.1 and forthcoming in Windows 10. This can be problematic for specific use-cases where an organization would like to test an update with all of their apps to avoid software issues. Single-app mode (iOS) or Kiosk-mode (Android) can limit the user from performing an update.

EMM solutions can provide a way to enforce updates to ensure that vulnerabilities are patched. This can be performed through a security policy that blocks network access or other enforcements to encourage users to perform the update. But there are obstacles and a lack of APIs (Application Program Interface) to enforce the mobile operating system updates from the EMM.

WEARABLES

Wearables and smartwatches didn't become a topical risk concern for most organizations until the release of the Apple Watch. There was certainly the fear of the unknown. Are these devices risky or not? What happens if a device is hacked or lost? The fact is that wearables and smartwatches have been around for years prior to the Apple Watch, and some can be paired with an iOS device in addition to Android and Windows devices.

There are fundamental differences between wearables and smartwatches versus their mobile device counterparts. These smartwatches typically require a pairing app on the mobile device to allow the smartwatch to be paired over the air. The most important difference of a smartwatch versus a mobile device is that the built-in security for smartwatches is more proximity based rather than PIN or passcodebased. With mobile devices typically the first security requirement most organizations have is to enforce a PIN or passcode on the device to protect the data in the event that the device is lost or stolen. Smartwatches typically use a proximity-based approach. This can rely on identification of it residing on the user's wrist, and when it's removed a PIN or Passcode prompt is enabled to protect it. In other smart-watches, this proximity-based protection is based on whether the device is communicating over Bluetooth to the paired mobile device. When the Bluetooth connectivity is lost, the PIN or passcode is enabled.[2]

Management APIs are starting to appear for the Apple Watch. Apple has provided the ability to detect when an Apple Watch has been paired to an Apple iPhone. Other controls include blocking access to enterprise data using containerization as well as blocking the smart-watch pair apps. Look for this area to mature over the next few years. For now, considering using App-level security or containerization to mitigate the syncing of enterprise data to smartwatches. In the case of the Apple Watch, there are methods of embracing the Watch Kit exten-sion for those enterprise apps that you would like to sync with a smart-watch, and level the encryption capabilities in combination with this.

DEVICE ENCRYPTION AND CONTAINERS

Most of the devices today across iOS, Windows, Android, and more provide operating system-level encryption either enabled by default or as an option. Furthermore, this can be enforced by the EMM as part of the enforcement policy. This is one of the fundamental requirements of most mobile security strategies.

But encryption alone doesn't prevent users from sharing data. To accomplish that requires a container to control sharing of corporate data through separate encryption and data loss prevention controls. This container can include email, secure access to corporate content (fileshares), web browsing, and corporate apps and data. Data can be shared across the apps within the container, but can block unwanted cloud services or sharing of data with apps outside of the container. In addition, it should provide controls to copy/paste, open-in, sharing, and other behaviors that allow moving of data to and from the corpo-rate container.

[2]https://www.mobileiron.com/en/whitepaper/smartwatches-wearables-and-mobile-enterprise-security MobileIron Analysis of Smartwatch Security Risks to Enterprise Data

Typically the container is separately encrypted from the rest of the device. This autonomous encryption can prevent the container data from being exposed, even if the device is compromised or infested by malware. And furthermore, when a device is compromised the container can wipe the container data in real-time. People frequently ask about targeting data in memory on a mobile device. Aside from some device specific vulnerabilities, most device compromises require jailbreak or rooting behaviors, which additionally require a reboot of the device. Therefore to complete the compromise, you reboot the device thus wiping volatile memory. So previously viewed documents are gone, and not exposed to memory analysis tools such as IDA Pro, *after the compromise*. There are always exceptions to every scenario, so it's important to embrace the other outlined layered security to eliminate any single point of exposure. In this case, app reputation, mobile threat prevention, requiring mobile operating system updates and patches before accessing corporate data, operating system compromise detections, quarantine, and numerous other controls can further protect against these types of exposures.

PINs, PASSWORDS, AND PASSCODES

Determining passcode enforcement policy can be challenging for some organizations. It typically stems from traditional PC and Server 8-character password policies that require various complexities to achieve compliance or traditional security best practices. This is a prime example of traditional policies that just don't work well in the mobile world. Requiring a user to enter an 8-character complex password to unlock their mobile device makes for a horrible user-experience.

Users are accustomed to a 4-character PIN. Most EMM policies can then enforce various complexities or wipe a device after 10 bad PIN entries. Many security conscious organizations have embraced App-level or Container-level passcodes to protect corporate data. And in those cases, some have incorporated a 6-character PIN or passcode at an App-level or Container-level.

Bottomline: it comes down to the organization, but it's very important to consider the broader mobile security controls not found in the typical PC world (eg, Wipe after 10 bad passcode entries). It's important to balance that with the user-experience to avoid lack of mobile adoption or causing users to circumvent security controls in other ways, commonly referred to as Shadow IT. Some of these options can include fingerprint authentication through Apple's Touch ID or Samsung's fingerprint scanner. This can be use to authenticate at a device or container level.

CLOUD

One of the key questions most people ask is how can an organization separate personal cloud from enterprise cloud (Enterprise File and Sync Share services). Early on, mobile administrators would blacklist the personal cloud apps, but this is like playing "whack-a-mole." If you block one personal cloud repository, the users will just find another.

At a device level it's important to provide an enterprise solution to users. Some of the most popular solutions have an Enterprise version of their app, which can also embed the SDK provided by an EMM. This allows that app to then work in unison with the EMM containerization to require users to upload enterprise data (in the container) using only that app versus personal cloud apps. Another approach is to leverage a containerized documentation collaboration app that allows webdav access to the enterprise cloud repository. For additional tips, see the "File-Level Security" section in this chapter.

FILE-LEVEL SECURITY

Users want to store documents and files in personal cloud services. In many cases they don't distinguish between personal and corporate files; therefore it's common for an employee to upload a file to share with another employee, business partner, or prospective client. Mobile Data Loss Prevention (DLP) controls and containerization are designed to prohibit such behaviors to avoid mobile data loss. But when these controls ruin the user-experience, employees will attempt to circumvent those controls resulting in Shadow IT. To overcome this issue, another approach is to embrace the personal cloud services rather than block them.

File-level security is about tying security to a corporate document. With this approach, a user can use their favorite cloud service for uploading and sharing corporate documents. When a file is shared to a personal cloud service, the file is first encrypted before being uploaded. If the file is then shared with another employee, the key escrow at their company allows the file to be downloaded and decrypted for the employee to access and use the document. But when the file is shared with a nonemployee, the file remains encrypted and unusable by the nonemployee. This is a nice compliment to a defense-in-depth mobile strategy and creates a great user-experience for mobile users.

SUMMARY

Threats will always exist and continue to evolve. Implementing a layered security approach is key to succeeding in your mobile security strategy. But success is not always about avoiding a breach altogether, but also being prepared to respond to it. A thorough incident response plan can mitigate data loss and prepare you for when a breach occurs. If your security team doesn't have a mobile-specific incident response methodology, they should. Engage your team to ensure the vetted processes are in-place to response. Many times we've heard "it's not a matter of if, but a matter of when"; be prepared. All of the recommendations outlined should also have a tie-in to your incident response plan. The countermeasures defined in the chapter should help in implementing your defense-in-depth mobile security strategy.

CHAPTER 4

Ensuring Mobile Compliance

Mobile devices are making their way not only into the Enterprise through BYOD and Corporate Issued scenarios, but also in hospitals, retail stores, traveling technicians, logistics, and many other industries. Due to the sensitivity of the data on the mobile device, ensuring security as well as compliance is important to many organizations. But as previously mentioned, mobile devices architecturally are designed to be very different from legacy PCs and Servers, therefore the traditional security policies and legacy compliance requirements do not always apply. This also impacts compliance security approaches and requirements as well. Let's explore some of the most common regulatory, industry, and government compliances.

PCI

More retailers are using mobile devices in their stores to improve the customer experience, provide "line-busting" during heavy periods, and better security in light of all of the retail breaches. But some of the PCI requirements don't map entirely to the design differences of mobile device operating systems. A good example of one key difference is anti-malware, as outlined in chapters "Understanding Mobile Data Loss Threats" and "Mobile Security Countermeasures." Again, anti-virus alone on mobile can identify threats, but there are many limitations to mitigating the threat. Therefore, EMM or MDM is required to respond to the threat with a quarantine. These differences in mobile are what prompted the release of the Mobile Payment Acceptance Security Guidelines, designed for mobile devices running Point-of-Sale (POS).

As of this writing, PCI DSS 3.1 standards are now in-place (July 1, 2015). To support Mobile POS, the PCI Council also released the PCI Mobile Payment Acceptance Security Guidelines, v1.1 in July, 2014. Furthermore, Mastercard has set a deadline for the retailers to support

Table 4.1 PCI DSS 3.1 Requirements — Summary for Mobile

Section	Requirement
2.2	Hardening—Configure system (mobile device) security parameters to prevent misuse
2.3	Encrypt all nonconsole (remote) administrative access using strong cryptography
4.1.1	Facilitate strong authentication for Wi-Fi by pushing certificates
8.3	Facilitate two-factor authentication for remote access to CDE
10.5.4	Audit logging of mobile device activity on device
12.3	Usage policies and procedures for tablets & PDAs

EMV credit cards on the POS devices by October, 2015,[1] or incur the liability of a breach after that deadline. It's important to note that the PCI Council considers EMV an important move toward deterring future breaches, but has also stated that EMV "is not a silver bullet" and device management is also recommended, which is where Enterprise Mobility Management solutions come in. All of these requirements are intended to further thwart the numerous retail breaches over the last 2 years.

In the context of using mobile devices for POS, there are some important differences when compared to a traditional POS cash register such as a Windows-Embedded POS. Mobile devices offer a better defense-in-depth solution to the many malware threats that have impacted legacy POS. Let's further explore the PCI requirements in the context of Mobile POS.

Ensuring PCI compliance for your Mobile POS devices requires that the merchant review two groups of requirements and guidelines. The first is the PCI DSS 3.1 requirements. It's important to note that most of the requirements apply more broadly to the overall requirements, for example the CDE (Cardholder Data Environment). So one must extract the mobile-specific requirements. The following Table 4.1 summarizes these.

Digital certificates are an important part of the mobile-specific PCI requirements for administrative access and Mobile POS Wi-Fi access. Fortunately, digital certificates are very easy to deploy on mobile devices; iOS, Android, Windows Phone, and others fundamentally support certificates. With the mobile device under MDM/EMM management allows for configuration profiles to be pushed down to

[1] http://www.mastercardadvisors.com/_assets/pdf/emv_us_aquirers.pdf.

Table 4.2 PCI Mobile Payment Acceptance Security Guidelines — Summary

Checklist	Guideline
	Prevent account data from being intercepted when entered into a mobile device
	Secure distribution of account data
	Secure access to & storage of account data
	Controls over account data while in use (preventing copy/paste, screen shots, file sharing, etc.)
	Prevention of unintentional or side-channel data leakage
	Prevent account data from interception upon transmission out of the mobile device.
	Deploy MDM or MAM, protect device against malware
	Prevent unauthorized logical device access (secure lock screen, full disk encryption, etc.)
	Create server-side controls and report unauthorized access
	Prevent escalation of privileges (detect root or jailbreak and Quarantine) + Offline detection & quarantine
	Remotely disable payment application
	Harden supporting operating systems (restrictions)

the device. This can simultaneously include the creation and deployment of a certificate to the device for uses such as secure access to the Wi-Fi, VPNs, Application Tunnels, secure web access, and more.

MDM/EMM also allows for hardening of the mobile device so that the administrator can disable unnecessary services for a Mobile POS. This may include disabling the Bluetooth, IRDA, SD Card, AirDrop, microphone, and much more. DLP controls can also be used to disable copy/paste, screenshot, open-in, etc. It's important to review these options for your chosen Mobile POS device type whether it's iOS, Android, Windows, or some other device; as these features vary from device to device. In addition, these features also change with operating system upgrades as new features are released. Your MDM/EMM can allow you to stay ahead of these new features and disable them upon upgrades of the Mobile POS devices.

The other set of requirements are more guidelines, but very important for adapting your PCI security strategy when incorporating mobile devices into the POS mix. The Mobile Payment Acceptance Security Guidelines outline the following (Table 4.2):

Probably the single largest fundamental difference in meeting the requirements for Mobile versus a PC-based POS is how anti-malware functions. In mobile, anti-virus or anti-malware is just another app on the device. We've become accustomed to Anti-Virus simply removing the threat on a PC once detected. But on a mobile device, the application sandboxing prohibits the anti-virus from removing a threat since it's isolated just like any other app. In the mobile device world a new approach has emerged referred to as an App Reputation Service or Mobile Threat Prevention. Fundamentally, an App Reputation Service leverages the EMM's app inventory and compares that to its inventory of apps to identify apps that contain malware or risky behaviors that may expose data. Additionally, some may or may not use an anti-virus-like app on the device. These are more tuned to mobile threats, but most importantly are still tied to the MDM/EMM. When a threat is identified, these solutions communicate the threat to the EMM and the EMM will quarantine the device. The quarantine can automatically remove the POS App and/or its data from the device upon detection. Additionally, the quarantine can optionally use a mobile gateway to automatically block the device's connectivity on the network to protect the infected device from impacting the broader CDE.

Another major threat is a mobile device compromise stemming from jailbreak (iOS) and rooting activity (Android). There is a plethora of techniques for jailbreaking or rooting a device. In addition to the variants of jailbreaking techniques, there are tools meant to hide the fact that the device has been jailbroken. For Android, there are also a variety of ways of compromising a device. These can stem from an Android device-specific vulnerability, side-loading/side-jacking, use of the ADB and USB controls, Custom ROMs, and much more. When this occurs the operating system sandboxing is circumvented and security is weakened on the device, opening up the device to a variety of data loss threats and attacks.

Quarantine options can automate the response to a mobile device compromise ranging from a Full Wipe of the device, to a Selective Wipe where just the POS data and/or apps are removed from the device. Note that these behaviors vary slightly across the platforms including iOS, Android, Windows, and more.

One concern for retailers is the scenario of a lost or stolen device. When this occurs it's no longer on the retail Wi-Fi and thus the retailer loses visibility and management of the device. Furthermore, if an attacker then tries to target the device to steal credit card information, the EMM can't send a quarantine command to the device to wipe the credit card data from the Mobile POS App. Knowing this drawback, some EMM solutions have added offline operating system compromise detections and the ability for a local policy on the device. This allows the data to be selectively wiped from the device when it's in an off-the-network state and allows this to be performed more in real-time. PCI embraced this by adding it to the Mobile Payment Acceptance Security Guidelines in version 1.1.

HIPAA

In healthcare, mobile devices offer a more cost-effective solution to mobilize healthcare employees within a hospital. These devices are typically much cheaper than proprietary traditional mobile devices and can be more easily updated through simple app updates over-the-air, rather than a full device update through a tethered approach which can be cumbersome. Also, mobile devices are being used to improve the patient recovery by offering a temporary mobile device to patients to use while they're in the hospital recovering. Even more interesting is the fact that in-home healthcare is making a dramatic comeback as nurses and physicians are now equipped with mobile devices to provide in-home healthcare and the fact that this historic approach to health-care is becoming popular again.

The Health Insurance Portability and Accountability Act (HIPAA) outlines Privacy, Security, and Enforcement Rules for health information. This encompasses the HITECH Act outlining the rule for Beach Notification.[2]

There are many categories that comprise these standards, but the one most applicable to securing Patient Health Information (PHI) on mobile devices is the 164.312 Technical Safeguards to protect Electronic Patient Health Information (ePHI). This includes Confidentiality, Integrity, and Availability (CIA) of all ePHI. The CIA security objectives model is outlined by NIST.

The 164.312 Technical Safeguards outline an overall strategy for securing patient health information, which can be applied to mobile devices. The following Table 4.3 outlines the safeguards.

[2]Federal Register, Friday, January 25, 2013 Department of Health and Human Services Vol. 78, No. 17.

Table 4.3 HIPAA 164.312 Technical Safeguards

Section	Requirement
Access Control	164.312(a)(2)(i)—Unique User Identification 164.312(a)(2)(iii)—Automatic Logoff 164.312(a)(2)(iv)—Encryption and Decryption 164.312(a)(2)(ii)—Emergency Access Procedure (written procedures)
Audit Controls	164.312(b)—Audit Controls
Integrity	164.312(c)(1)—Integrity 164.312(c)(2)—Authenticate ePHI
Authentication	164.312(d)—Person or Entity Authentication
Transmission Security	164.312(e)(1)—Transmission Security 164.312(e)(2)(i)—Integrity Controls 164.312(e)(2)(ii)—Encryption

Mobile is well-equipped to meet these requirements. The ease of support for certificates, as well as the fundamental application sandboxing, provides a great strategy for protecting data-at-rest and data-in-motion.

Containerization is commonly supported on mobile devices today to allow separation of health data from personal or nonsensitive data. This containerization provides encryption for data-at-rest including email and attachments, application data, and content (eg, health records). Additionally, data-in-motion is secured through an encrypted tunnel, for example an application tunnel or per-App VPN (iOS) without the need for a VPN.

Certificates can be used for authentication and automatically pushed down to the device for authenticating email, fileshares, content access, applications, and secure web access. This inherently provides integrity of the device connecting as the device can be validated as a sanctioned device. And closed-loop compliance actions provide an automated way of quarantining a device that falls out of compliance by selectively wiping the health data from the device.

CJIS

The Criminal Justice Information Services (CJIS) Security Policy outlines minimum security requirements for information protection. It applies to agencies that access "Federal Bureau of Investigation (FBI)

Table 4.4 CJIS Mobile-Specific Security Requirements

Section	Requirement
5.5.7.3.1(1)	Apply available critical patches and upgrades to the operating system as soon as they become available for the device and after necessary testing as described in Section 5.10.4.1
5.5.7.3.1(2)	Are Configured for local device authentication
5.5.7.3.1(3)	Use advanced authentication
5.5.7.3.1(4)	Encrypt all CJI residents on the device
5.5.7.3.1(5)	Erase cached information when session is terminated
5.5.7.3.1(6)	Employ personal firewalls or run a MDM system that facilitates the ability to provide firewall services from the agency level.
5.5.7.3.1(7)	Employ anti-virus software or run a MDM system that facilitates the ability to provide anti-virus services from the agency level.
5.5.7.3.3(1)	CJI is only transferred between CJI authorized applications and storage areas of the device.
5.5.7.3.3(2)	Remote Locking of device Remote Wiping of device Setting and locking device configuration Detection of Rooted and Jailbroken devices Enforce folder or disk-level encryption
5.10.1.2	2 Encrypt all CJI resident on the device. Minimum 128-bit encryption, FIPS 140-2 certified, PKI

Criminal Justice Information Services (CJIS) Division systems and information to protect and safeguard Criminal Justice Information protection."[3] (Table 4.4)

Many of these requirements overlap with other regulatory and industry compliances. It's important to note that the anti-virus requirements can be fulfilled through App Reputation or Mobile Threat Prevention solutions that integrate with MDM (and EMM). Again, anti-virus alone on mobile can identify threats, but there are many limitations to mitigating the threat. Therefore those solutions that integrate with MDM/EMM provide much better closed-loop actions for mitigating threats by quarantining the device.

SUMMARY

Many more regulatory and industry compliances exist, but they largely overlap in terms of securing data-at-rest and data-in-motion. What's important is to incorporate the approaches in mobile that differ from

[3]https://www.fbi.gov/about-us/cjis/cjis-security-policy-resource-center/view.

traditional security approaches in the PC world. What has been provided should provide a good baseline. But regulatory and industry compliances change over time, sometimes annually or bi-annually. Be sure to stay up-to-date on these requirements through the respective governance websites and publications. This will assist in avoiding over-sights and possible fines. In the case of PCI, always consult your PCI QSA (Qualified Security Assessor) to ensure up-to-date compliance.

CHAPTER 5

Developing Your Mobile Device Security Strategy

When developing a mobile device security strategy one should embrace both security *and the users.* Without embracing the users and making security as invisible as possible, users may become frustrated and abandon the solution. Additionally, it may cause "Shadow IT" causing users to find ways to bypass security controls. This is a different mindset to traditional security approaches, and should be incorporated into any mobile security deployment to ensure success. It's not about restrictions, but about enablement.

PROACTIVE CONTROLS

Any holistic security strategy should include proactive, reactive, and live monitoring controls. Proactive controls should protect the data-at-rest and the data-in-motion. Fundamentally, a device PIN/Password and encryption are important, but for a security conscious organization, further DLP controls are required. For example, if a user receives a corporate email with an attachment, there's nothing preventing an employee from opening the attachment and uploading it to a cloud service, sharing it with nonemployees, and more. This is where separation of personal and enterprise data becomes important.

An encrypted container for enterprise data provides proactive protections from the personal persona on the device to avoid comingling of the data and data loss of enterprise data. Through both encryption and DLP controls, the organization can control enterprise data, while leaving the user's personal data alone. This will provide protections against accidental or intentional sharing of enterprise data with cloud services, other email accounts, copy/paste, screenshot, and more. Additionally, this container provides a level of protection against malware downloaded outside the container, as the container is encrypted and controlled separately from the rest of the device. While although this anti-malware approach proactively protects against malware, the reactive approaches outlined should be used as well.

For data-in-motion, secure tunnels are recommended. While many organizations use VPNs on the mobile devices as a carryover from their PC and laptop management, application tunnels or container-specific micro-VPNs provide arguably better security in a mobile deployment. Full device VPNs typically allow all apps on the mobile device to access enterprise private network. This is a concern because it may inadvertently allow malware to infest the enterprise network behind the firewall.

A fundamental approach across many of the mobile operating systems is a per-App VPN or Application Tunnel. This allows the administrator to be more selective in terms of what apps are allowed to access to enterprise network, and thus all other apps are blocked including malicious apps.

Strong authentication in mobile is typically delivered using Certificates. Digital Certificates have made a comeback with mobile devices. These devices fundamentally support certificates, which can be used for strong authentication to enterprise resources. This helps ease deployments as an enterprise can avoid passwords that may change over time and cause confusion amongst users and result in helpdesk calls. Certificates can be pushed down with configuration profile (eg, corporate email), and allow the user to securely access email. For those organizations concerned about storing certificates on the device, certificates can be stored in the container for access to sensitive services. More broadly, identity and access management may be a consideration for your organization.

Organizations also have concerns about users traveling and accessing the enterprise network over an insecure network, such as the Open Wi-Fi at a local coffee shop. The user could become a victim of an interception or MitM attack. These attacks can target both unencrypted as well as encrypted data. Using the outlined controls of both per-App VPN and certificate-based authentication can provide proactive controls against mitigating these threats by providing confidentiality and integrity controls.

REACTIVE CONTROLS AND PROTECTIONS

Reactive controls should be real-time or near-real-time.

The breaches from 2013 to 2015 enumerated that it typically takes days, months, even years to identify a data breach.[1] By today's standards and expectations, this is nowhere near responsive enough. This delay in response has led to exposures of large amounts of credit cards, patient health information, and personnel records. Mobile devices and Enterprise Mobility Management afford us the ability to more quickly identify threats and automatically respond to those threats.

It should be apparent at this point that by far the biggest threat is from malware and operating system compromises. In mobile, this can stem from a variety of vectors including, but not limited to:

- User Jailbreaking an iOS device and loading an app outside of the App Store
- User Rooting an Android device and side-loading an app outside of Google Play
- Attacker distributing an app through a malicious email link or SMS message
- Developer who unknowingly builds an app using a third-party SDK unaware that the SDK includes malware or risky behaviors, and posts it in the App Store or Google Play
- Malicious user who circumvents a curated app store and the app vetting security processes and posts an app for download

The EMM's mobile device client compromise detections, App Reputation or Mobile Threat Prevention, and quarantine are good deterrents to mobile malware. This will ensure the integrity of your devices, and allow the EMM to respond to threats by quarantining devices when they fall out of compliance. This can allow the EMM to:

- Perform a full wipe of a device (Best for Corporate Issued Devices)
- Perform a selective wipe by just removing the corporate data and/or Apps (Best for BYOD)
- Block the device's network access to enterprise resources by using a Secure Mobile Gateway and/or Network Access Control
- Alert, log, and report on out-of-compliance devices

It's important to note that this may require the EMM to wait until the next time a device checks into the console, perhaps up to 4 hours. While that is far quicker than the aforementioned breaches, for some

http://www.verizonenterprise.com/DBIR/2015/

organizations this still isn't quick enough. As a result, some EMM solutions now allow a local policy to reside on the device to respond more immediately to an identified threat, perhaps even if the device is not network connected and cannot communicate back to the EMM management console. This can allow the corporate container on the device to be selectively wiped to avoid a breach of corporate data. This occurs more in real-time without the need to "phone-home" to the console.

These proactive and reactive mobile security controls can be summarized in the following table:

Holistic Mobile Security

Proactive	Reactive
• PIN/Passcode • Encryption • Strong Auth/Certificates • Containerize Corp Content & Apps • Per-App VPN • Secure Mobile Gateway • Network Access Control • User or device certificate to thwart MITM attacks • Vulnerability Scanning	• Malicious & Risky App Detection • Jailbreak/root detection • App Reputation/MTP • Closed-loop compliance actions • Auto-block enterprise network access • Selective Wipe (Corp Apps, Data, Email, etc.) • Compliance Reporting/ Alerting

This mobile security strategy should also be complimented with Live Monitoring. Live Monitoring can include a console that provides a view into the changing landscape of security posture of devices and their access to enterprise resources on the network or in the cloud. It should also send alerts to key staff members, provide logging and audit trails, and include integrations with security information and event management (SIEM) and big data analytic tools. This enables the day-to-day activities for maintaining the overall health of the mobile deployment, but also feeds into the incident response plan, and even perhaps post-mortem forensics.

Not all incidents warrant a forensics investigation. In fact, most security incidents are a matter of daily security administration, much of which is automated through your EMM. Additionally, these may be an indication that a security control or access was overlooked and should be refined.

It's important to note that the EMM console will commonly alert on both network and device threats. A few notables that may require further investigation can include:

An unregistered device attempting to access restricted resources (internal or cloud) using a legitimate account. Authentication failures may be an indication of brute force attacks, integrity failures, or MitM attacks.

- A device that has been compromised. This presents a risk to the enterprise data. Ensure that the mobile device has been automatically quarantined, and follow-up with user as part of your incident response processes.
- Detected malware on a mobile device. Determine the variant of the app with the malware, or the malware itself to make a determination if it stemmed from a compromised device, a malicious link, or some other attack vector. After examining, decide if the security controls should be modified. Also determine if any other devices exhibited the same malware to understand the extent of the infestation.

EMM is a nice compliment to a forensics investigation. As these mobile devices are becoming inherently more secure, performing forensics on data-at-rest on a mobile device is becoming increasingly difficult. Many mobile device forensics products require the PIN to gain access to the device to allow a forensic image to be created. Additionally, other approaches require a resident vulnerability be used to load a custom RAMdisk, jailbreak or root the device, access DFU mode, amongst other approaches. All of these create a hurdle to imaging the device and performing analysis. It's for these reasons that EMM is becoming a more important part of liturgical and nonliturgical forensic investigations.

MOBILE DATA LOSS THREATS AND COUNTERMEASURES FLOW

If we model the flow of the attack vectors, we can gain insight into layers of defense that encompass proactive, reactive, and live monitoring controls. Whether we have a mobile device connecting to the Wi-Fi locally, or connecting remotely to target the network, we can take a

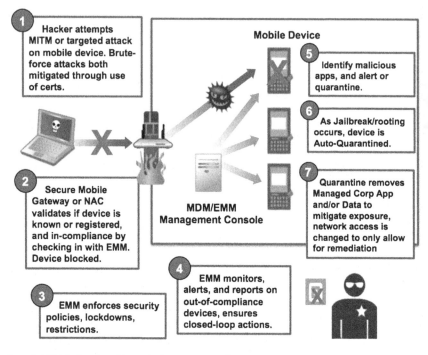

Figure 5.1 Mobile Data Loss Threats & Countermeasures Flow.

very similar approach to mitigating data loss. The following diagram outlines the flow (Figure 5.1):

This diagram leverages many automated controls that include a Secure Mobile Gateway or Network Access Control that, from a network perspective, control access to resources including email, content repositories, intranet websites, and application servers.

More and more organizations are also embracing the cloud. EMM can also provide integrations with cloud services to blacklist a mobile device and/or user to prevent further access. Additionally, the selective wipe locally on the device can remove the downloaded cloud data and credentials, thus revoking access.

MOBILE DLP METHODOLOGY

I've covered a lot threats and countermeasures in this book. With the mobile threat landscape evolving so quickly it's important to assemble an actionable methodology to securing mobile devices, apps, content,

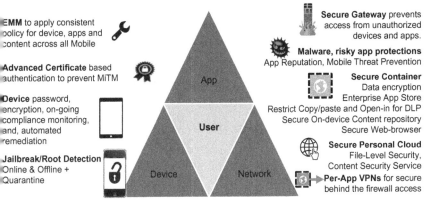

EMM to apply consistent policy for device, apps and content across all Mobile

Advanced Certificate based authentication to prevent MiTM

Device password, encryption, on-going compliance monitoring, and, automated remediation

Jailbreak/Root Detection Online & Offline + Quarantine

App
User
Device   Network

Secure Gateway prevents access from unauthorized devices and apps.

Malware, risky app protections App Reputation, Mobile Threat Prevention

Secure Container Data encryption Enterprise App Store Restrict Copy/paste and Open-in for DLP Secure On-device Content repository Secure Web-browser

Secure Personal Cloud File-Level Security, Content Security Service Per-App VPNs for secure behind the firewall access

Figure 5.2 Mobile Device Threats Countermeasures Framework.

and network access. The following outlines triad approach for implementing a secure mobile device and network strategy (Figure 5.2):

FUTURES

As of the writing of this book; wearables, Internet of Things (IoT), and more are emerging. I've outlined research which highlights the risks of smartwatches and wearables to mobile enterprise data. Although if we consider enablement, an organization can further mobilize employees with wearables and smartwatches in a secure manner that embraces securing the data-at-rest and data-in-motion using many of the strategies outlined in this book.

As your mobile journey progresses, it will be important to further understand the benefits of extending these security approaches to these wearable devices that are an extension of our mobile devices. Set up a lab, test these devices, get more familiar with them, and understand their behaviors.

MOBILE SECURITY REQUIRES A NEW APPROACH

Mobile devices are here to stay and they're quickly becoming the primary device for the new generation of workforce. But mobile also moves at a very brisk pace like we've never seen in IT. Taking a quote from my mentor 'develop an iterative security framework, noting that it will become outdated quickly, it must be a model constantly updating, therefore the security team needs to work differently, because *Agility is the new Security.*"

Printed in the United States
By Bookmasters